MEERKATS

MEERKATS

PHOTOGRAPHY BY NIGEL DENNIS
TEXT BY DAVID MACDONALD

NH
NEW
HOLLAND

First published in 1999 by
New Holland Publishers
London • Cape Town • Sydney • Auckland

24 Nutford Place, London W1H 6DQ, United Kingdom
80 McKenzie Street, Cape Town 8001, South Africa
14 Aquatic Drive, Frenchs Forest, NSW 2086, Australia
218 Lake Road, Northcote, Auckland, New Zealand

ISBN 1 86872 310 0

Project Manager: Pippa Parker
Editor: Sue Matthews
Designer: Dominic Robson
Cover design: Janice Evans
Copy Editor: Simon Pooley
Proof Reader: Gill Gordon

Reproduction by Hirt and Carter Cape (Pty) Ltd
Printed and Bound in India by Ajanta Offset

2 4 6 8 10 9 7 5 3 1

Half title: A meerkat on sentry duty.
Full title: The meerkats must survive in the harsh
environment of the Kalahari Desert.
Right: Two meerkat kittens are watched over by an elder.
Overleaf: A meerkat scans the horizon for predators.

ACKNOWLEDGEMENTS

The Photographer

I wish to thank South African National Parks for their kind support and assistance during my stay in the Kalahari Gemsbok National Park. The photographs in this book were taken under the guidance and control of South African National Parks. In particular I would like to thank Dries and Henriette Engelbrecht, Giel and Emmerentia de Kock and Tinnie and Lettie Visser for their friendship and kindness. I am very grateful to Professor Tim Clutton-Brock and the Nossob Suricate Research Team, and also wildlife film makers Lynn and Philip Richardson for invaluable help with my meerkat photography. As always my wife Wendy accompanied me on all of the Kalahari visits. Many of the meerkat photographs in this book were taken on foot, often right at ground level, making the photographer vulnerable to attack by large predators. I am extremely grateful to Wendy for selflessly keeping watch for danger from the roof of our vehicle during the long hours I spent photographing. I wish to dedicate this book to her.

NIGEL DENNIS

The Author

This account is based on the better part of a decade of research in the Kalahari Gemsbok National Park, and I warmly appreciate the hospitality and support during this time of both the National Parks Board and Pretoria University's Mammal Research Institute. I also thank my various friends and collaborators, notably Sean Doolan, Mike Knight and Gus Mills, and, in the preparation of this text, Lauren Harrington.

DAVID MACDONALD

MEERKATS

As the sun comes up over the Kalahari, a small creature disentangles himself from the surrounding huddle of bodies and pokes his head out of his underground burrow. Seeing no sign of danger, he clambers out and stands on his hind legs, facing the sun and feeling its warmth spread through his cold body. A moment later he is joined by other members of the group, stiffly taking up their positions around the burrow, but ever alert to the hazards that exist above ground. This is a group of meerkats, their mutually dependent relationships some of the most intriguing of mammalian societies.

Meerkats (*Suricata suricatta*) are known by a confusing array of common names: suricates, slender-tailed or grey meerkats, and stokstertjies (Afrikaans for 'stick-tails'). In fact, the name 'meerkat' derives from the Afrikaans for marsh cat, although they are not cats at all, and certainly do not occur in wet, marshy environments. Rather, they are mongooses, a family of carnivores known as the Herpestidae, and are found in the arid and semi-arid lands of southern Africa, in areas where annual rainfall averages less than 600 millimetres (23.6 inches). Part of the meerkats' range, which extends from southern Angola, across Namibia and Botswana and into the northern parts of South Africa, overlaps with those of the yellow mongoose, the larger Cape grey mongoose, the slender mongoose and, to the north, the banded mongoose.

Even when standing upright on their hind legs, supported by their stick-like tails, meerkats are little more than 30 centimetres (12 inches) tall. Yet their fang-like canines, small beady eyes and long curved claws impart the appropriately fierce appearance of carnivores. About 236 species belonging to the order Carnivora exist

Meerkats are mongooses, and inhabit the arid and semi-arid western parts of southern Africa.

8

today, but many more have gone extinct in the past. Modern carnivores are extremely diverse – comprising a range of species including the polar bear, badger, African wild dog, giant panda, raccoon, palm civet, river otter and the Indian tiger – all descended from a common ancestor that lived about 60 million years ago. The only hallmark of that shared past is a set of scissor-like back teeth, called the carnassials, designed to shear through flesh, and even these are now unrecognisable in a few species which have special diets, for instance the termite-eating aardwolves.

Yellow mongooses live in similar habitats to meerkats, but are more carnivorous, more commonly hunt alone and live in smaller, more loose-knit groups.

The diversity of the carnivores is largely a reflection of their adaptations to secure food – from the flesh-devouring lynx to the clam-eating sea otter, the fruit-guzzling kinkajou to the bamboo-crunching giant panda – and a corresponding variety of social lives. While the African palm civet leads a solitary, nocturnal existence, feeding on small vertebrates, insects and fruit that are too small to share, members of a lion pride in the Serengeti feast together on a wildebeest, and two dozen wolves may combine forces as a highly skilled, co-ordinated pack to tackle a moose.

THE MONGOOSES

Today's mongooses are similar to the earliest fossil ancestors of modern carnivores and have many ancient anatomical characteristics. Indeed, their skeletal morphology and dentition have withstood the test of time and have remained virtually unchanged for 40 to 50 million years.

The first mongooses probably originated in Eurasia about 50 million years ago and migrated to Africa during the early Miocene, 20 to 30 million years later. Perhaps they followed the migration of grazing ungulates, feasting on a tasty supply of insects in islands of dung. But unlike the cats, which had similar origins, mongooses did not cross the Bering land-bridge

and thus didn't make it to the Americas. Until, that is, humans took the small Indian mongoose to the New World, to Jamaica, in 1872 to control rats in sugarcane plantations. Like so many other such introductions, this plan backfired, and now these little mongooses thrive in the islands of Jamaica, Hawaii, Puerto Rico, Cuba and the Dominican Republic, threatening the local wildlife and acting as carriers of rabies. The rest of the mongoose family is restricted to the Old World, but within it species are found as far afield as southern Europe, Israel, Egypt, Asia and southern Africa.

Mongooses weigh about a kilogram (a little over 2 pounds), and are much like the weasels and mink of the northern hemisphere in appearance. However, 8 of the 36 species of mongoose are flamboyantly sociable, a feature that radically distinguishes them from their weaselly counterparts. These gregarious species, which include the meerkat, exist as cohesive groups bound together by mutual dependency. All forage by day for insects, a plentiful food source in the southern climes they inhabit. At least two other species, the yellow mongoose of southern Africa and the narrow-striped mongoose of Madagascar, are collaborative in the chores of infant care, but generally forage alone. Mongooses that hunt alone or in pairs, such as the slender mongoose (the most widespread mongoose of sub-Saharan Africa), feed on larger but less abundant prey such as small rodents, birds or lizards.

The sociable species tend to live in open areas, while the solitary species usually live in dense forested habitat and are often nocturnal, which reduces the benefit of many individuals looking out for predators. Mongoose lifestyles are fairly adaptable, though, as demonstrated by the Egyptian mongoose, which is abundant throughout Africa and around the Mediterranean into Israel. It is normally rather solitary, active mostly around dawn and dusk, and has a mixed diet, but where hotels provide a ready source of food it forages with others and will even breed communally. In Spain, Egyptian mongooses even hunt for rabbits as a 'family'. This ability to adapt their lifestyles to suit circumstances is seen not just in one species in different places, but also in single individuals at different times. The narrow-striped mongoose, for example, forages alone in the dry forests of Madagascar (mostly for grubs), but hunts in co-operative groups in the wet season for mouse lemurs and chameleons.

Most carnivores generally do not form gregarious groups, but among those that do, a complete spectrum of sociality exists, from pairs (usually a male and female), to large stable groups that share a home range or a food source, to highly co-ordinated groups that hunt together. It is the unveiling of the intricacies of these vibrant social relationships that has made the study of carnivore behaviour such a thrilling and adventurous field of scientific enquiry during the past thirty years.

THE SOCIABLE MEERKAT

Meerkats are close to the pinnacle of carnivore sociality. A meerkat group is typically made up of some ten individuals, but groups of as many as 30 members are not exceptional. Each group may consist of one to three breeding females, up to four adult males (one of which is usually dominant over the others), several yearlings and as many as a dozen of the year's offspring. In contrast with those societies that are essentially extended families – a breeding pair and their offspring – meerkat groups comprise several adults of each sex. Individuals may live together for much, if not all, of their lives, although both yearlings and adults of both sexes regularly disperse and join other groups. These details were effectively unknown as recently as the early 1980s, but subsequent research has revealed that meerkat society is among the most intriguing of mammalian societies, with the potential to reveal profound lessons about the evolution of co-operation among all mammals, including ourselves.

As miniature carnivores, no bigger than the tassel on a lion's tail, meerkats face the uncomfortable predicament of being both predator and prey. On the one hand they are

As representatives of the cat, hyaena and dog families, respectively, the lion (top), brown hyaena (above left) and black-backed jackal (above right) are all distant kin of meerkats. All belong to the biological order Carnivora.

rapacious hunters, searching every hole and crevice for signs of beetle grubs and geckos, and digging feverishly at the scent of food, their heads obscured by a cascade of flying sand. On the other hand, they risk being snatched by swooping birds of prey and must constantly scan the skies for signs of danger. The open savanna habitat where they reside is also home to several snakes and more than 60 other carnivores, from lions and spotted hyaenas to black-backed jackals. All of these are a potential threat to meerkats and their mongoose cousins as they forage throughout the day.

While a single meerkat would not be able to perform either the foraging or the vigilance job satisfactorily, as a member of a team it can share the burden of looking out for threatening predators, so that each individual need pause from foraging only momentarily to scan the horizon. In this way, at least one meerkat will always be on the lookout, and the larger the group the less frequently an individual need interrupt its feeding – a case of many hands (or eyes) making light work.

A more sophisticated security arrangement is one in which a sentinel is appointed with full responsibility for lookout duties, while the rest of the group feeds undisturbed. From a guard post with a good view of the surrounding barren landscape – high up in an acacia bush, at a prominent position on the skeleton of a dead tree, or on top of a large termite mound – the sentinel utters regular peeping sounds. This 'watchman's song' signifies to the rest of the group that he or she is on duty and alert. All members of

A meerkat stands sentinel duty, allowing the group to concentrate on foraging.

the group, with the exception of the kittens, will take a turn at sentinel duty, although the time they spend on duty varies greatly between individuals. While some members appear to work tirelessly as guards, others seem to be freeloaders. At some age or stage, however, all of the members of the mob end up pulling their weight in a corporate endeavour. In general, the older subordinate males tend to shoulder much of the burden of group security, and often appear obsessively devoted to their duties. They seek the highest vantage points, up to 4 metres (13 feet) high, and often stay on guard duty for an hour or more. Dominant males, in contrast, only rarely act as sentinels and even then for no more than a ten-minute period. Adult female meerkats also do guard duty but usually for shorter periods than the subordinate males and from less ambitious vantage points, while young adults of both sexes guard the least.

Sentinels have a repertoire of calls for different levels of danger. They 'peep' in response to a predator that is far away or not particularly threatening, but voice a trilling 'growl' if the predator continues to approach or a dangerous species appears. The closer or more dangerous the predator, the more guttural the guard's alarm call. White-backed vultures, pale chanting goshawks and bateleur eagles rarely attack meerkats and so are only worthy of a brief alarm call, whereas tawny and martial eagles, which do attack them, both elicit a long and agitated call. Meerkats can even distinguish a martial eagle from a black-breasted snake eagle, a species that looks very similar but is not considered a threat. Smaller raptors, such as kestrels or pygmy falcons, are tolerated to within about 30 metres (32 yards), with

Meerkats fear tawny eagles, and can identify them at prodigious distances.

the meerkats only showing consternation when they approach any closer. Yet the sight of a martial eagle more than 300 metres (328 yards) away causes an immediate response, and all members of the group dash to the nearest den. Jackals are even more terrifying to meerkats, and often appear to taunt them by lingering close by, pinning down a growling and trilling group for hours on end as precious foraging time ebbs away.

Adult meerkats are especially vigilant when young, vulnerable kittens emerge from the den. At the sight of a lanner falcon they will fling themselves in a protective blanket over their charges, even though this raptor is not normally considered a threat. Newly mobile kittens are a harrowing responsibility for their guardians and, as they stumble about on their first excursions, much of the day may be spent carrying them to the safety of the nearest bolt hole. Young meerkats have much to learn about vigilance, although certain aspects appear to be instinctive. A related species, the dwarf mongoose, has been studied in this regard. From the age of 32 to 40 days old, the young emit warning calls in response to raptors overhead, and by five or six months they have a strong compulsion to climb onto high vantage points. Apart from appointing guards, dwarf mongooses have developed a remarkable partnership with hornbills. In the course of their daily foraging the mongooses flush out prey not normally available to the hornbills, which in return make warning calls if they spot a raptor that preys on the mongooses, even if it poses no threat to themselves. This partnership is so refined that the hornbills wait for the mongooses to emerge from their den in the morning before beginning their day's foraging.

The price that sentinels pay for guarding would seem to be considerable: not only do they give up their own foraging

Meerkats encounter yellow cobras all too frequently in the dark confines of their burrows, but they may be immune to this snake's venom.

time, they are also subjected to scorching temperatures while exposed as a beacon to passing predators. On average, a subordinate male guarding a medium-sized group loses one-fifth of his foraging time, but for smaller groups this may be even higher. More importantly, although the sentinel is continually watching for predators, his post may be some distance from the foraging group members. If an attack occurs, he has to run several metres to rejoin them, making himself an easy target.

At first sight, then, it may seem that the guards are heroic figures, but another interpretation for this apparent altruism is that once an animal has eaten its fill, the safest place to be is on guard. With eyes peeled and concentration focused, the guard will be the first to spot danger and quickest to take evasive action – so perhaps its motives are entirely self-centred. The jury is still out on which interpretation is correct, and both may turn out to apply under different circumstances.

Despite their tiny stature, meerkats do not take the risk of predation meekly, but go out of their way to harass at least the less daunting of their predators. What they lack in size they more than compensate for in aggression and force of numbers. This mobbing behaviour may be an attempt to banish predators; for example, meerkats devote much enthusiasm to harassing yellow cobras in the vicinity of favoured burrows, which may reduce the risk of future underground encounters in dark, tight corners. The cobras, which often grow to more than a metre (3.3 feet) long, defend themselves with repeated strikes, which the meerkats deftly dodge while nipping at the snake's tail. Part of the meerkats' bravado can be attributed to their ability to withstand snakebites. They are reputed to be able to tolerate six times the venom dose that would kill a rabbit.

Goshawks and other smaller raptors are attacked on sight, a mob of meerkats running underneath the bird and biting at its undercarriage as it takes off. The small Cape fox is also scorned by these diminutive demons, and regularly chased from its daytime lair, but more formidable opponents, such as jackals and honey badgers, are avoided. However, if cornered, the group will confront an animal as large as a black-backed jackal. With tails raised high, backs arched and fur fluffed out, the meerkats bunch together and move as one seemingly giant attacker towards the startled foe, with mouths agape and hissing.

Such bravado is, of course, somewhat dependent on the size of the group. A band of at least six meerkats will mob a jackal, whereas smaller groups and individuals are more inclined to flee to a refuge. Here a dangerous predator can be harangued from the safety of the den – it is extremely amusing to see a jackal with its head down one entrance to a burrow and nether regions inelegantly aloft, while a mob of meerkats growls abuse from an adjoining entrance.

Meerkats and other gregarious mongooses have a reputation for heroic rescue attempts when one of their number is imperilled. The most famous case recorded was a group of banded mongooses climbing a tree to rescue one of their companions from the talons of a martial eagle! Meerkats are also unusual in that they will care for a group member that is injured, whether it is the result of a failed predatory attempt or a skirmish with a neighbouring rival. The group not only escorts the invalid back to the safety of the den, but tends and feeds it until it either recovers or dies. Adults huddle with the injured individual, even if unrelated to it, and restrict their own daily movements, remaining close to the den and providing the invalid with prey (often large preferred prey items such as geckos and lizards). While the injured member is weak and unable to travel far, the whole party travels less, moves more slowly and returns to the den earlier than normal. Furthermore, an invalid may be greeted more often in the low creep ordinarily reserved for dominant individuals, and groomed even though it is unable to reciprocate. In general, there is much vocalisation, huddling and greeting, all centred around the invalid, during convalescence.

A TERRITORIAL SPECIES

In the southern Kalahari, groups of meerkats inhabit expansive territories of between 200 and 1 000 hectares (approximately 500 to 2 500 acres), littered with dozens of bolt holes and dens. This ensures that a meerkat is seldom more than 50 metres (55 yards) from the safety of a burrow, the exact location of each apparently etched into its memory.

Apart from providing havens from predators, dens and burrows are used as overnight shelters from the cold and for siestas during the heat of the day. They are most often excavated in the firm soils of dry riverbeds and the edges of pans, rather than in the intervening dunes, where digging is a wearisome battle against soft, unstable sand. Dens and burrows are usually either dug by the meerkats themselves or shared with other species such as ground squirrels or yellow mongooses. Occasionally, a sheltered rock crevice may be used.

Border patrols and general territory maintenance are an essential part of meerkat life. The group works together to maintain the bolt holes and dens, digging and passing sand backwards along an industrious chain gang. Individuals will scent-mark

Previous pages: A meerkat family, vigilant as ever.
Below: Lizards, such as this agamid, are rich pickings compared to the meerkat's main diet of beetle grubs.

frequently near the territory border, and after detecting the scent of a neighbouring group. Some groups are so meticulously tidy that all members defecate into a communal dung pit to avoid fouling their own territory.

Territories are defended rigorously against neighbouring competitors and gang wars are not uncommon. A typical meerkat 'skirmish' typically starts with two groups encountering one another at the border. The opposing groups take up their positions, with all individuals standing on their hind legs, jostling for position, jeering at the enemy and leaping higher and higher in the air. A slow approach culminates in a charge, often with the dominant male in the lead. The larger mob usually wins and the victors indulge in an explosion of sociality, grooming and hugging each other, wiping their odorous anal pouches across each other's flanks and finally defecating en masse in pits feverishly dug on the battleground.

FORAGING HABITS

Harvesting prey within the vast foraging ground is carried out in a painstakingly methodical manner: a small patch is worked each day, with scarcely a nook or cranny left uninvestigated.

Generally, a different area is visited each day and it may be a week or more before a particular route is repeated. This probably allows the food supply to be replenished, preventing the meerkats from eating themselves out of house and home.

Meerkats feed mostly on insects (mainly larvae and adult beetles), which make up more than three-quarters of their prey in the Kalahari. However, they do not hesitate to take more varied food items when the opportunity arises.

In addition to their ability to withstand snake venom, meerkats also seem to be immune to scorpion venom. Certainly, a sting that could kill a child apparently causes them little more than mild discomfort. After nonchalantly flicking a scorpion from under a log, a meerkat will dodge the prey's pincers with lightning speed and lunge repeatedly, even directing its bite at the sting, until the scorpion is weakened and can be eaten. Two different scorpion species are a regular feature on the meerkats' menu. *Opisthophthalmus*, which subdues its prey with large pincers rather than powerful venom, is eaten from the rear end, the pincers dangling like a Chinese moustache from the meerkat's jaws. In contrast, *Parabuthus*, which has smaller pincers but an ominously large abdominal sting, is eaten headfirst.

While foraging, meerkats do not so much walk as shuffle – incessantly pawing and shifting sand with their forepaws as they feel for crevices hiding prey.

Above: A scorpion's pincers dangle, like Chinese moustaches, from a meerkat's jaws.
Overleaf: Baby-sitting kittens so their mother can feed.

Meerkats

Millipedes, which are the staple diet of banded mongooses, are eaten by meerkats only when their more preferred prey is unobtainable. Indeed, it appears that meerkats find millipedes distasteful, as they will repeatedly roll a millipede over in the sand as they back away, shaking their heads and wiping their mouths, after biting it.

Small reptiles are an important food for meerkats, and extensive burrowing is undertaken to unearth reptiles such as nocturnal barking geckos from their subterranean refuges. In fact, meerkats tend to eat more reptiles, including snakes and agamids, than any other mongoose species.

Meerkats do not take large, active rodents, although these are caught and eaten by the yellow and banded mongooses in southern Africa. Yellow mongooses are the same size as meerkats so it is unlikely that meerkats are incapable of catching these rodents – although this does require a somewhat different strategy to that used for insect foraging. While yellow mongooses forage alone and silently, meerkats forage as a group uttering a constant 'brrp' call, apparently to keep in contact with one another. This may be another reason why meerkats do not catch the larger rodents, as this constant calling is enough to scare off most small mammals, which are in any case thinner on the ground than insects and geckos.

Insects are often found in large numbers in a small space, such as a dung pile or an ant nest, so meerkats are frequently seen feeding side by side. A neighbour that gets too close will be rebuked, but overt squabbling over food is uncommon. When larger prey is on the menu, meerkats may even co-operate. The sausage-sized gecko *Chondrodactylus angulifer* lives in deep burrows, and two or sometimes three group members will need to work together for half an hour of hard digging to catch one. This kind of co-operation usually pays off as two of these geckos are often found in one burrow.

Different parts of the meerkats' home range may harbour different types of prey. In the Kalahari, for example, grubs of tenebrionid beetles tend to be especially numerous in the red sand dunes, whereas barking geckos and large scarabid beetles prevail in the tangled roots of rhygozum bushes. Large, orange cockchafer beetles are found in the riverbeds.

The fruitfulness of certain areas is also dependent on weather conditions, since the hotter and wetter it is, the more beetle grubs there are. As the moisture in the sand retreats deeper and deeper, so do the beetle larvae, until they are out of reach of even the most determined meerkat excavator. Fortunately, however, meerkats are opportunistic feeders and are never entirely reliant on a single food source – a prudent strategy when food availability is dependent on the highly variable rainfall of southern Africa's arid lands.

The dry winter is a hard time for the meerkats, as food is in short supply and competition between individuals is high. The larger larvae have descended deeper into the soil, and food tends to be limited to ants and small insects. The meerkats must simply survive on less food; foraging for longer is not an option for a small skinny animal in the cold evening temperatures, and meerkats never take the risk of travelling by night, when predators lurk under cover of darkness.

Digging for grubs is hard work, given that a meerkat regularly digs through its own weight of sand in a few seconds. After such energetic activity, a midday siesta is a well-deserved break, and there is still time to have some fun. Regardless of status or age, everyone participates, joining in a display of gymnastic tumbling, sparring with imaginary enemies and grabbing one another's tails. Such uninhibited play is a demonstration of the strength of cohesion between group members, especially since hierarchy appears momentarily immaterial.

REPRODUCTION

As the breeding season approaches, the importance of hierarchy becomes apparent: breeding, in all but the most productive years, is generally a privilege of the dominant male and female. In meerkat society, an individual's social status tends to increase with age, although females may attain breeding status within a group by inheritance, or by the outcome of aggressive competition. Hostility between two older females within the group can be particularly intense, with the dominant female initiating a lengthy period of targeted aggression towards her competitor. Indeed, most of the aggression within a group is between older, same-sex individuals.

Birth and First Weeks

Females produce between one and three litters each year, dominant females usually producing more litters than subordinates. In meerkats, three to seven kittens make up a litter, whereas the less social slender and yellow mongooses have only one or two kittens in each litter. Most kittens are born between January and March (although the breeding season extends from October to June), in large breeding dens that become the centre of the group's activity for a short time after the birth of the litter. Ordinarily, a group departs soon after emerging from the den in the morning, on a long and arduous foraging excursion that covers several kilometres and keeps them away from the den all day. Immediately after the birth of the young, however, the group emerges from the den later in the morning than usual, spends more time basking in the sunlight, greeting, grooming and scent-marking, and only ventures forth on short foraging sorties of less than 100 metres (109 yards).

Over the next few weeks the young become the focus of intense social activity. To the onlooker, the meerkats' regular visits underground and out of sight are tantalising when all that can be heard are the excited social vocalisations emanating from the burrow's entrance.

The kittens remain in the den for their first four weeks, during which time they are entirely dependent on milk. For the next few weeks they are still nursed, but also begin eating chewed-up food brought to the den by the other adults in the group. Most of this provisioning is done by adult males – yearling and juvenile helpers will also contribute but are more likely to 'cheat', snatching the food back as soon as they have given it up, particularly when times are tough and there is a shortage of food.

Above: Grooming is a key to meerkat social status; here a subordinate animal grooms a superior.
Opposite: At the end of a long day's foraging the meerkat group congregates around its den.

While the kittens are still restricted to the breeding den, other members of the group assume baby-sitting duties, allowing the mother to forage. Lactation uses up a lot of a female's energy resources, and the mother must find enough food for herself to ensure that she can provide milk for her brood. All members of the group, both male and female, do their turn, although the most regular baby-sitters are often previous offspring of the mother. Baby-sitting is a punishing task as it involves an entire day of fasting for an animal used to eating every few minutes! In fact, regular baby-sitters may lose as much as 10 per cent of their weight in the course of rearing a litter.

For the most part, baby-sitting is a solitary activity, but as many as three baby-sitters may remain at the den at any one time. Whereas adults usually baby-sit alone, yearlings are often accompanied by additional group members, raising the possibility that they are trusted less. Certainly, adults are invariably more assiduous in kitten care than are yearlings. In a group of baby-sitters, each can take a turn at foraging nearby, but a single baby-sitter never leaves the entrance of the den, and can only snatch a few moments to forage when the rest of the group returns at the end of the day. Although all members of the group participate to some extent, breeders (especially dominant ones) are less likely to baby-sit than non-breeders, and some mothers never take a turn. In a small group, however, there is increased pressure on all individuals to take on a share of these chores.

Baby-sitters are also responsible for deterring and harassing predators in the vicinity of the den, and for promptly taking the kittens below ground when danger looms. Small predators such as yellow mongooses, as well as solitary meerkats and

Despite their marvellous adaptations, much about the meerkats' fate is determined by their harsh environment. Rainfall may secure the survival of kittens by nurturing their food supply, but a torrential storm may doom them in a flooded burrow.

neighbouring groups, are generally chased away. Yet the tables are sometimes turned when a neighbouring group stumbles upon a guarded litter during a foraging trip. The meerkats will often attempt to drive the baby-sitter from the den, and if successful the kittens are invariably killed. Fortunately, many of the most diligent baby-sitters are the heavyweights of the group, with a 'bodyguard' physique that may prove a great asset.

The hazards for a very young meerkat are numerous, but among the most serious is the risk of flooding. Meerkats breed during the rainy season to ensure a plentiful supply of beetle grubs to feed the litter, but heavy thunderstorms can cause flash floods in the pans, with devastating consequences. Baby-sitters have been seen to make heroic rescues, single-handedly moving

entire litters from the flooded breeding den to safety in another den in higher ground. However, the brutal reality is that all the baby-sitters in the world cannot save a chilled litter in a burrow which is awash with water.

Perhaps the zenith of care-giving in meerkats is achieved by wet-nurses – females that nurse kittens other than their own (technically known as allo-nursing). Most wet-nurses are subordinates that have lost their own litters but, remarkably, some females lactate without showing any outward sign of pregnancy. Nursing another female's offspring is, in fact, common among carnivores, and the phenomenon of spontaneous lactation is also seen in dwarf mongooses, Ethiopian wolves and, at least anecdotally, in several other canids. This may explain how females can successfully adopt very young kittens that have been orphaned. Sharing the burden of nursing might not only increase the chances of survival of the young by ensuring a steady milk supply, but also increase their mother's chances of survival, since lactation imposes such great stresses that it may threaten her own future.

The greatest risk period for kittens is between the ages of three and five weeks, when their fur is sparse and they chill rapidly. Indeed, most mortalities of young kittens coincide with rapid temperature declines during heavy rainstorms and cold fronts. The kittens probably spend most of their first few weeks huddling together and with their baby-sitter to keep warm, continuing this behaviour even after they have emerged from the den. When left by the baby-sitter above ground they huddle together, but if they lose contact with one another or a chilling breeze blows up, they will call loudly until rescued.

Huddling not only enables meerkat kittens to withstand cold, but also conserves more of their energy to be used for growth. A growth spurt at this crucial age is important as the kittens must be able to keep up with the group on long foraging trips when they are only eight weeks old (dwarf mongoose kittens may have to travel at only 25 days old). Other carnivores have lifestyles that enable them to extend this safe period for the young. The semi-gregarious yellow mongoose, for example, forages independently for small mammals, yet all the group members provision the young at the den for up to ten weeks. For meerkats, transporting insects to

Meerkat sociality is complex, and involves the forging of close social bonds that may last a lifetime.

the den for this length of time is simply too inefficient, and might lead to prey in the immediate vicinity being depleted while the rest of their established territory remains unpatrolled.

Emergence from the Den

Meerkat kittens first emerge from the den at about five weeks old. Slightly more than two-thirds of them are likely to survive to foraging independence at about 12 weeks of age, but during the intervening period they are extremely vulnerable to predation. At about eight weeks, when the kittens first accompany foraging parties, they sit and wait to be fed, although they often rush towards a calling adult or the sound of other kittens being fed. On these initial foraging trips the kittens may be guarded in a crèche near a vantage point, and close to a bolt hole, while the foragers search for food nearby. The same members of the group that undertook the most baby-sitting duties at the den now take responsibility for crèche-guarding duties.

By the second week after they emerge from the den, the kittens are each adopted by an adult as an 'apprentice'. They shadow their 'tutor' constantly, fiercely defending it from other kittens and begging loudly for food. The tutor, relenting to the apprentice's demands, donates much of its best-quality prey, handing over rare succulent items such as scorpions. Nevertheless, in the scrum of begging, feeding and squabbling between kittens, each apprentice will frequently change tutor. By the third week away from the den the kittens have gained confidence and will snatch food not being proffered. At this stage adults may provide incapacitated but still living prey to the kittens, even repeatedly returning the item when the youngster fumbles the kill.

At the tender age of four months a young meerkat is expected to be able to feed itself. However, juveniles up to seven months old may try to solicit food from animals that had previously fed them, even though these animals may by then have some responsibility for provisioning a new litter of kittens.

Despite the great diligence of meerkats in the care of their young, the survival of the kittens appears to depend mainly on environmental conditions. The number of litters each female bears during a given year is largely influenced by the amount of rainfall, rather than by the number of helpers at hand to share the task of kitten-care. The same appears to apply to kitten survival – it may be that, as long as there is the minimum necessary number of helpers required in the group, any further increase in helpers is of little benefit to the young.

Rainfall also determines the abundance of insect food that is available for the young kittens when they are weaned, as well as being a major influence on whether a female conceives. Meerkats appear to take the opportunity to breed whenever it has rained, although rain at conception does not necessarily mean rain at the birth, so the risk exists that conditions may be poor by the time

Above: The Auob River in the Kalahari is normally a dry bed of silt, but heavy wet season showers provide water that will foster a crop of beetle larvae to sustain the next generation of meerkats.

the mother is lactating and the kittens are weaned. A shortage of food may even make lactation impossible. As a result, litters are sometimes abandoned during droughts, and cessation of early rains may mean no kittens are successfully raised that year. When food is abundant, however, meerkat females can quickly recover from the demands of producing and weaning one litter and come into oestrus again, extending the breeding season into the early winter months during particularly wet years.

How much rain falls is clearly crucial to meerkats, but when it falls, and how it is distributed, is also of vital importance. In a year when all rainfall occurs between October and December, animals breeding later in the season are facing extremely poor conditions by the time the kittens are born 60 days later. If the same amount of rain falls in another year, but evenly distributed between October and March, moderate conditions will prevail when the kittens are born and may even allow early breeders to raise two litters. In order to cope with the double stress of lactation and pregnancy concurrently, however, the female needs to increase her food intake significantly – in this case, a fair amount of assistance with the tasks of guarding and provisioning her young would seem essential.

Factors other than rainfall also determine when meerkats mate, but these remain unknown, adding to the mystery of their reproductive affairs. For instance, why is it that neighbouring

Overleaf: Meerkats are physiologically adapted to withstand intense heat. The down-side of this is that they lose their body heat quickly, and must spend time soaking up the warmth of the morning sun before setting forth on a day's foraging.

groups do not necessarily breed at the same time, whereas births within groups are close to simultaneous? The answer probably lies in some kind of social stimulus, a theme common to certain other social carnivores, including the dwarf mongoose.

Infanticide

A rather unappealing aspect of many mammalian societies is infanticide – the most widely known case is in lion prides, where it is chiefly the males that are responsible. This grisly practice is probably also a feature of the private lives of meerkats, but in their case the evidence points to females as the guilty parties. The incentive for male infanticide is to end investment in the offspring of rivals, but within a meerkat group males may not be sure which of them fathered the litter, as subordinate males probably sneak at least some matings. An apparent contradiction of this line of thought is that bands of males taking over a neighbouring group and ousting the resident males have been known to assist in raising young conceived before the event as their own. These are the exact circumstances in which incoming male lions kill their predecessors' progeny. A possible explanation for these divergent responses to identical circumstances is that because meerkats face so many chores best achieved by large groups (guard duty, baby-sitting, fighting rival groups), even unrelated recruits are welcome.

It appears then, that female-inflicted infanticide is the dark secret of meerkat society. In most cases, a dominant pregnant female kills a subordinate female's litter, ensuring that all food brought by the helpers to the den is available for her own kittens. Yet there is little evidence that the offspring of dominant females actually do fare better as a result of her murderous actions.

Pregnant subordinate females are also often forcibly expelled from the group in the latter stages of pregnancy, which may partly serve to remove any females likely to be tempted into infanticide. These ostracised mothers are allowed to rejoin the group only after they have given birth to their doomed litter. Once allowed back into the group, they apparently neither kill the dominant female's kittens nor become pregnant again for a period of at least several months, but are likely to take on the role of wet-nurse.

In most cases reproduction in subordinate females is suppressed hormonally, although exactly how this is achieved by meerkats is not fully understood. In dwarf mongooses, the female sex hormone oestrogen, which is essential for successful reproduction, never reaches a level sufficient to trigger ovulation in subordinate females. This is less marked in older females, and some do breed successfully.

In contrast, male dwarf mongooses have high levels of sex hormones regardless of their rank, but subordinate males are either physically prevented from mating by the dominant male, or are only 'allowed' to mate with subordinate, infertile females. Nonetheless, subordinate males keep their options open by retaining their fertility and mating opportunistically. Like meerkats, dwarf mongooses may live in very large groups, and under these circumstances the dominant male may be unable to guard the breeding females at all times. This increases the probability of a subordinate being able to sneak in some surreptitious mating.

Why do subordinates put up with such treatment? In both dwarf mongooses and meerkats the dominant male fathers most of the offspring, so the likelihood that the females in a group are related to one another is quite high. For a non-breeding subordinate female the forces of kin selection, essentially an investment in relatives to ensure that at least some of her own genes will be represented in future generations, may make her more inclined to help raise the litters of others. Even if the dominant female suppresses her subordinates' reproduction, kinship may provide sufficient incentive for the latter to stay. In banded mongooses, where the females in a group are not closely related because multiple males breed with several oestrus females, subordinates may need the added incentive of being allowed to raise their own young to stay in the group and behave as stakeholders.

DISPERSAL

The adolescents and subordinates of a meerkat group may have a long wait before they attain breeding status, and sometimes young males run out of patience, opting instead to leave the group and resort to brute force. A roving group of bachelors may attack the males of a neighbouring group, steal the territory, and with it the resident females. Young females may be just as tempted

Meerkats

to disperse, and occasionally they elope en masse, deserting their companions to join a group of bachelor males. Both sexes tend to leave in the company of relatives, particularly siblings or group members that they helped to rear.

On the whole, fewer females disperse than males, and they are more likely to do so alone. The females also appear rather discerning in the selection of a new group, so they not only spend more time alone, but also travel further than do males, making them more susceptible to predation. Lone females, however, are more likely to be accepted into a new group than are lone males, possibly because such female immigrants are often exceptionally diligent co-operators, paying their way, so to speak.

Meerkats of both sexes improve their chances of successfully dispersing by carrying out prospecting forays before they make the final move. By doing so they become more familiar with the area beyond the borders of their home territory, allowing them to reconnoitre new dens and vantage points. Investigation of scent stations may also help an individual to fathom the local social scene before deciding whether emigration is a sensible option. A young male joining a group with many breeding males may be a poor prospect, whereas seeking out a group well endowed with females may be worth the wait. Most emigrations by mongooses are to groups with few older individuals of the same sex.

Adult meerkats generally disperse early in the breeding season when the most mating opportunities are available. The males often leave the group when the dominant female is pregnant, perhaps because they are a lot more likely to encounter female outcasts in this period, with which they can set up their own groups. It certainly also makes sense for them to leave

Previous page: Meerkats sparring.
Above: The tsama melon is a source of water for some desert mammals, such as brown hyaenas, but meerkats hardly ever drink. They get liquid from their prey.

their own group well before the baby-sitting duties commence!

Of course, not all dispersal is voluntary – some individuals may be forcibly evicted from the group. Apart from the dominant female evicting subordinate females that are pregnant or in oestrus, senior males may be ousted by a rival meerkat gang during a take-over.

Dispersing or ousted individuals setting up a new group are at an enormous disadvantage, because the division of labour that makes meerkat society so successful becomes problematic when groups are very small. In a group with only three or four members, the pressure of work is worsened because lactating females are automatically excused from both guard duty and baby-sitting. This means that the work schedule for the remainder of the group may verge on being unsustainable. The logical outcome is skimping on guard duty, with potentially fatal consequences. Such 'short-staffed' groups are extremely vulnerable to predation.

ADAPTATIONS FOR LIFE IN THE DESERT

Being both pint-sized predator and succulent prey presents dilemmas aplenty for the meerkat. These pressures are intensified because the drama of meerkat life is played out on the arid, hostile stage of southern Africa's Kalahari desert. Here rainfall is erratic and temperatures fluctuate widely, both through the day and between the seasons – with the hot, wet months of October to April merging into a cold, dry winter between May and September. Summer temperatures are generally above 20°C (68°F), and often reach 35°C (95°F), or even extremes of 40°C (104°F), which result in sand temperatures exceeding 70°C (158°F) and the difference between sun and shade being as much as 30°C (86°F). Beneath the clear skies of winter, temperatures can fall nearly 30°C (86°F) from a warm midday high to a bitterly cold nighttime low of -14°C (20°F).

Virtually all of the year's rain (less than 300 millimetres, or 12 inches) falls in the summer months – mostly between January and April – when flowering plants rush to germinate, carpeting the desert with colourful flowers. Rain typically falls during short thunderstorms in the late afternoon and early evening. During the dry winter, however, rainfall is negligible. The days are fairly warm, although maximum temperatures remain below 20°C (68°F), but the nights are bitterly cold, with temperatures averaging 0 to 5°C (32 to 41°F) and plummeting as low as -10°C (14°F) or even -14°C (20°F), causing regular ground frosts. For most of the year winds are moderate, although dust devils frequently whirl across the landscape. In August and September, however, violent winds from the north and west cause major dust storms and batter any trees in their way.

In these harsh conditions, meerkats are trapped in an eternal energy crisis. Losing time to cold winter mornings and roasting summer noons, they struggle to find sufficient hours in the day in which to forage. Luckily they are physiologically adapted to cope with periods of low food supply, and with the desert's extreme heat.

Meerkats have a remark-ably low metabolic rate – 58 per cent of the usual level for an animal of their size. One advantage of such a low metabolism is that internal heat production is minim-ised, a handy feature in the sweltering conditions of the Kalahari in midsummer. The reduced risk of overheating may allow meerkats to forage for longer, at hotter tempera-tures, while the lower energy demands lessen the need to

After the rains yellow thornflowers, *Tribulus terrestris*, spring up, signalling a time of plenty for the meerkats.

forage. An alternative tactic, typically adopted by rodents in the desert, is to shelter by day in cool burrows – but this is not an option open to the meerkat, with its diurnal lifestyle.

A low metabolic rate also limits evaporative water loss from the body, a crucial feature for desert dwellers. The meerkats' pelage, or coat, is an unusually good conductor of heat, enabling them to lose heat without concomitant loss of water. Even at temperatures as high as 40°C (104°F), meerkats can resist heating up for at least five hours by panting, like a dog, as a mechanism for evaporative heat loss. The larger mammals of the Kalahari take a different approach: eland and gazelle, for instance, allow their bodies to absorb heat gradually by day, but shed it at night.

The meerkats' capacity to lose heat easily, which is so useful in the heat of the day, does have a downside: they chill quickly at night. This requires stoking the internal fires with additional food to generate heat, and then cuddling up for shared warmth. Fortunately, burrow temperatures are far less variable than are the air temperatures above, so just as the meerkats' burrows remain cool havens under the midday sun, so they also provide warm shelters from the often bitterly cold night air of the desert.

Quite clearly, then, both the meerkats' behaviour and physiology equips these desert dwellers to survive in this harsh environment. Finding food, avoiding attack and keeping cool during the day-light hours are a constant strain. But back at the den at the end of the day, the meer-kats let down their guard and huddle together, some dozing, others grooming. In the safe warmth of their burrow, they fall asleep in an untidy heap from which protrudes a per-plexing array of legs, tails, paws and noses.

Previous pages: Meerkats, also known as suricates, are members of the order Carnivora. This makes them distant relatives of meat-eating animals such as lions, hyaenas, civets, wolves, bears, weasels and raccoons.

Above: In southern Africa's Kalahari Desert, the meerkat's world is subject to dramatic seasonal transformation. Following the rains, the sand becomes carpeted with colourful flowers, such as this vlei lily, *Nerina laticoma.*

A contented meerkat sits, relaxed but
watchful, amid blossoming yellow
thornflowers, *Tribulus terrestris.* The
rains have come to the Kalahari,
heralding a rare period of plenty for
these diminutive predators.

Meerkats

Opposite top: These bat-eared fox cubs will grow up, like the meerkats, to feed mainly on insects. Their staple diet will be termites. The fox cubs are cared for by both parents, which are believed to be monogamous.

Opposite bottom: Young meerkats, just old enough to accompany their group on foraging excursions, face many uncertainties on their early forays, and frequently clasp each other in anxious moments.

Above: The Kalahari is also home to a variety of other carnivores, each with its special adaptations to survive in the harsh environment. In the case of the cheetah, its speciality is speed; up to 100 km.p.h (62 m.p.h) in short bursts.

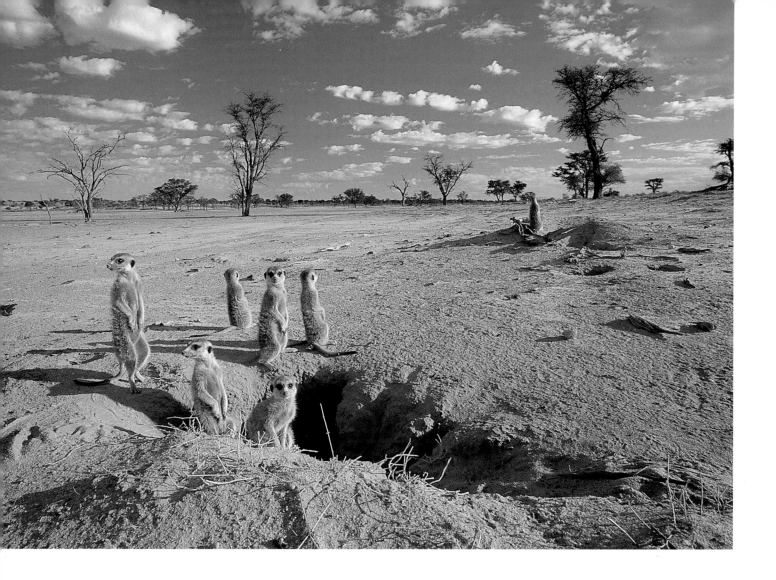

Above: Overnight, meerkats take refuge underground, where they huddle together for warmth. At dawn they emerge to face the sun, using its heat to restore their body temperatures before they set out on another long day's foraging.

Right: In the heat of the day, which may reach 40°C (104°F) in the Kalahari, meerkats may break off from foraging to take a siesta. Every so often one of the group will rouse itself to check for unexpected developments.

Above: Burrows are essential to the meerkat's lifestyle, whether for shelter from the elements or from predators. Meerkats appear to know the precise locations of all of the burrows within their sometimes extensive territories, so they can dash below ground at any time.

Left: Tunnels are an important asset to meerkats, and their entrances are frequently marked by the dominant male with a deft sideways swipe of the perineal glands under his tail.

Meerkats

Opposite top: The desert climate is far from uniform. During the dry season what little grass remains becomes brittle and yellow, and the parched, shifting sands throb with heat.

Above: As a rainstorm approaches, dust devils whirl in anticipation. The blinding clouds of sand create dangerous conditions for the meerkats – providing concealment for predators.

Left: After the rain, the dry riverbed of the Nossob valley in the Kalahari Gemsbok National Park is covered with puddles, but only very rarely, perhaps once or twice in a century, will there be sufficient rain for this ancient river to flow through the dry plains of the southwestern Kalahari Desert.

Meerkats

Life in a mob of meerkats involves a lot of work for the adults, but youngsters devote their energies to begging for food and boisterous play. Especially during siestas, while the adults doze away the rigours of a hard foraging session, youngsters fence and parry with their muzzles, clasp each other and roll over in rough-and-tumble fashion. Apart from having fun, the kittens are probably also cementing relationships that will determine the course of their adult social lives.

Meerkats

Left: The camelthorn acacia is a pivotal element of the arid Kalahari ecosystem. Antelope congregate in its shade and eat its seedpods, and the thick carpet of dung that develops beneath it can yield a rich harvest of beetle grubs for foraging meerkats. Sometimes, when the sun is blisteringly hot, meerkats will concentrate their foraging in these shade patches, charting a route from one camelthorn to the next.

Below: A meerkat stares piercingly at the horizon, its acute vision enabling it to spot meerkat rivals, terrestrial predators such as lion, hyaena and jackal, or to distinguish species of raptor such as the feared martial eagle, from hundreds of metres away. The tail provides the watchful meerkat with an extra prop, allowing it to take up a steady tripod stance that helps keep the target in sharp focus.

Meerkats

The great majority of meerkat snacks are invertebrates, mainly beetle grubs, but reptiles provide a succulent treat. Usually they are geckos only a few centimetres long, but sometimes more substantial reptiles such as snakes (above), or agamas (opposite and right), make a meerkat's day. Catching small geckos generally involves digging furiously at one entrance to their burrow, while nervously eyeing the neighbouring one in case the quarry tries to bolt.

Overleaf: Meerkats enjoying brief moments for reflection at the den after a hard day's foraging. Soon the desert will cool and they will retreat underground.

The springbok (left) is emblematic of the Kalahari, seeking shade beneath a camelthorn in the heat of the day (opposite top), and moving between the dunes and the dry riverbed (below) with the seasons. After the rains these antelope congregate in large herds to drop their young amid the first flush of succulent grasses.

Meerkats almost seem to ignore each other during their frenetic foraging activities, but back at the den for a midday siesta they often groom feverishly (opposite bottom). The pattern of who grooms whom hints at relationships within the group. For instance, a subordinate individual will approach a dominant one in a 'low creep', grooming first under its throat and then working its way around to the back of the neck (above and opposite top), a helpful place to choose as this is precisely where the recipient cannot reach. A solitary meerkat, estranged from its companions, will soon become infested with ticks on its neck and muzzle.

For a young meerkat the Kalahari landscape presents a great vista that it must come to know intimately. As an adult it will travel a huge home range, but will seldom be further than 50 metres (55 yards) from a bolt hole. Judging by the speed at which meerkats flee unerringly for shelter, they seem to remember the exact location of each hole.

Following the rains the red Kalahari sand is soon carpeted with vivid yellow thornflowers, *Tribulus terrestris*. This glorious luxuriance alerts the meerkat that the sand below will soon be rich in a post-rain glut of beetle larvae, not to mention more ambitious fare such as large scorpions (below). Meerkats appear to be largely immune to scorpion venom, experiencing only mild discomfort from bites that could be fatal for a small child. However, they rely on their speed and agility to avoid being stung if possible!

For meerkats, bolt holes mean survival, but the shifting sands are forever collapsing these life-saving sanctuaries. Day after day the meerkats toil to renovate their burrows, often working in a chain-gang of diggers to shift the sand backwards. As soon as a foraging band enters a new patch, its first priority is to renovate the local bolt holes.

Previous pages: Up on their tip-toes,
meerkats stand to attention as a foe
is spotted in the distance.

Below and Opposite: Sunset brings its
own hazards, as the desert rapidly cools.
The meerkats' small body size sentences
them to rapid heat loss, so the reward
of huddling together in the warmth
of the burrow is another important
benefit of group living.

Several small Kalahari carnivores compete for the same dens in which to raise their young. Two species of fox, the Cape fox (above) and the bat-eared fox (right), share this habitat with meerkats, and they are all hostile neighbours. Although a single meerkat is smaller than the flamboyant brush of the Cape fox's tail, their collective might and remorseless aggression enables the meerkats to triumph over both species.

Overleaf: With babies in the group, adult meerkats cannot afford to relax, and their vigilance is even more acute than normal. The babies are often unruly, but a warning growl from an adult reminds them of the seriousness of keeping a lookout for danger.

Meerkats

Meerkats

Previous pages: Nothing captures the sociability of meerkats more vividly than their practice of hugging their companions in moments of anxiety or relaxation. Among the most sociable of mammals, individuals may spend most, if not all, of their lives together with the same small group of companions.

Above: Even the meerkat's strong work ethic sometimes slips, as eyelids grow heavy, posture sags and the snout slowly droops. However, this momentary lapse in concentration by one individual is compensated for by the staunch vigilance of his companions. A benefit of group living is that many eyes make light work.

Backlit by the setting sun, two meerkats
share a rare tranquil moment. These
small carnivores are engaged in a
perpetual struggle against the elements,
contending with extremes of heat and
cold, and having to forage ceaselessly
for food in the dry plains of the
Kalahari Desert.

Strength of numbers protect meerkat groups from many encounters with predators, but some species strike terror into their hearts. They can spy a martial eagle (left) hundreds of metres away, and give a trilling warning that no group member will ignore. Rounding a bush and coming face to face with a black-backed jackal (opposite) will send a foraging band fleeing helter-skelter into a burrow. If a Cape cobra (below) happens to be snoozing inside, disaster ensues, but the meerkats take revenge when they come across the snake above ground. They will mob it remorselessly, nipping its tail and dodging its lightning strikes – a daredevil attack led by the dominant male.

More charismatic predators of the Kalahari, including the leopard (opposite) and the lion (above), have captivated the world's attention, but the minuscule meerkat is no less interesting and can look just as fierce (left).

Meerkats

Meerkats often encounter Kalahari ground squirrels (opposite top and bottom), and even cohabit with them occasionally. The squirrels appear like dim-witted country bumpkins alongside the sharp-witted meerkats, which may taunt and tease them if they have time on their hands. Nonetheless, meerkats will respond instantly (right) to a ground squirrel's alarm call.

Below: The yellow mongoose is among the meerkats' closest relatives in the Kalahari, but its social life is very different. Being more strictly carnivorous it tends to forage alone, although a group may share a den and co-operate in caring for the young.

Two kinds of Kalahari predator scan
the horizon: cheetahs are big enough to
concentrate on prey, but the pint-sized
meerkats hunt in a world populated by
giant foes and must be ever alert to this.

Meerkat babies are diligently tended by adults within the group, but pitted against the elements these kittens are all-too frail. Many perish from cold or drown in flash floods, forces of Nature against which even the most attentive adults can offer scant protection.

Overleaf: A sentry peers through the long grass, scanning for cunning predators on the hunt.

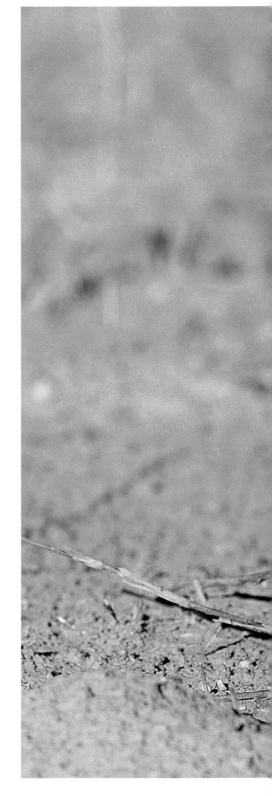

Meerkats

Previous pages: All manner of raised surfaces, from dead trees through the crowns of living trees and bushes to termitaria, serve as lookout posts for the ever-watchful meerkat appointed to sentinel duty. It is the subordinate males of the group who will perform most of the guarding, while the rest forage in relative safety.

Above: Sometimes more than one female within a group manages to rear young, and soon youngsters that were only recently begging for food will find themselves on baby-sitting duty.

Opposite: Young siblings stay close for moral support as they take a look at the intimidating world around them.

At sunset the dry grasses, camouflaged inhabitants and dusty skies of the arid Kalahari are tinged with golden hues.

Overleaf: During winter, when food is short and the nights are all-too long, the meerkats' morning warm-up period eats into vital foraging time.

Meerkats

Meerkats

Water is a luxury in the Kalahari and is only naturally available in temporary puddles after rain. However, most mammals are happy enough to drink at artificial bore holes pumped full by windmills. The meerkat, in stoic contrast, is rarely seen to drink at all, and obtains its water from the succulence of its prey.

Seeking comfort in an anxious moment,
a juvenile meerkat clings to an adult.
Meerkats are one of Africa's eight highly
sociable mongoose species.

Mating generally occurs in the safety of
the burrow. This dangerous liaison
above ground has the female scanning
the skies for predatory eagles.

95

Meerkats

Above: As darkness falls over the desert, a pair of lions watches and waits.

Opposite bottom: A meerkat heads for his sentry station. Is he sacrificing himself, going hungry while his companions feed, or has he eaten his fill and assumed guard duty as the safest way to spend his time?

Right: Gemsbok are a symbol of the Kalahari and namesake of the Kalahari Gemsbok National Park, which extends southward from Botswana into the northwestern reaches of South Africa. Here, the private lives of meerkats have been studied in depth.

Meerkats

The mothers of the carnivores of the
Kalahari teach their young to find food
and water, and in the meerkats' case, to
identify and distinguish between friends,
competitors and foes.

The midday siesta is a time for relaxing, socialising and scratching (above). At this time meerkats engage in a behaviour called hearth-rugging (opposite), which involves digging up a mound of cool sand and then spread-eagling onto it.

Overleaf: Detecting the scent of a rival gang as they arrive at a den, this mob, brandishing their stick-like tails, is ready to do battle and defend their territory. After hurling insults at the interlopers, they will charge them en masse.

Meerkats

Meerkats